STEAMBOATING ON THE MISSOURI RIVER IN THE SIXTIES

ROBT. E. LEE

STEAMBOATING ON THE MISSOURI RIVER IN THE SIXTIES

By
C.P. Deatherage

Ye Galleon Press
Fairfield, Washington
1998

This is a reprint of a little known historical booklet published in Kansas City over 70 years ago, describing early steam navigation of the Missouri River. Barges pushed by powerful diesel tugs still ply the mighty and sometimes capricious Missouri, but the glory days of steam are in the past.

ISBN 0-87770-677-8 (pb.)

Reprinted in 1998

Dedication

This little booklet is issued as a reminis-nt reminder to six friends, who learned look forward to the change from the enuous duties of city life to the care-e abandon of the "Out of Doors," and appreciate the murky, swirling, eddying aters of the "Big Muddy" as they made eir outing trips in their frail bark, the tle Stella, casting themselves adrift upon smooth bosom, knowing if they did not ock the boat," their voyage would be one unalloyed pleasure, a quiet and revivify g relaxation from the ever depressing, onotonous routine of their everyday life.

On May 11, 1878, they unloosed their ble at the Kansas City wharf, and after eventful and pleasant voyage tied up St. Louis May 18, 1878, and now forty- years after that memorable voyage this tle booklet is dedicated to GARDINER ATHROP, CORNELIUS MURPHY, and E. GUINOTTE, still with us, and to e memory of C. O. TICHENOR ALLACE LAWS and JUDGE JAMES IBSON, who have "Crossed the Bar," w resting on the tranquil waters of a undless eternity.

These six men made the trip in their at, Stella, whose picture we herein pro-ce as they tied up and rested at Hermann, o., therefore they were well acquainted th every peculiarity, the strange idio-ncrasies of the changeing currents of the Big Muddy," its smoky atmosphere at nrise; its refusal to mix with the waters of the Mississippi below its mouth, and knew how it hurled its mighty volume from bank to bank, from bend to bend in its mad rush to the sea, obeying gravity's unchangeing demands, and have witnessed the beauties of the river steamers as they ploughed their way upstream against the increasing volume of a June rise, and have been thrilled with the indescribable sight of the old-time steamboat rounding some distant bend with their stupendous symmetrical twin smoke stacks breathing out sparks from the fires below, bedecking the horizon after dusk with a myriad of sparkling stars, like a great volume of fireworks prematurely exploded on Memorial Hill.

Many other trips are accredited to these men in their annual outings, sometimes struggling against the mighty volume of an overflowing current on a June rise as far as Parkville on the Missouri, and sometimes as far as Turner on the Kaw. While to us none of them appear as heavyweights they assure us it took muscle and brawn to stem those currents, and as they floated leisurely on the down trips, their admiration for the beautiful scenery which all along the Missouri is unexcelled—even far exceeding the grandeur of the Palisades of the Hudson—south of the Gasconade.

May the survivors, their children and, their children's children, preserve this little booklet as a reminiscent memento of the achievements when the longest river on earth meant much to their forefathers, is the parting thought of the author.

C. P. D.

This booklet is a compilation of a series of articles written by C. P. Deatherage and published in June and July in The Kansas City News-Press.

Cornelius Murphy
James Gibson
J. E. Guinotte

Wallace Laws
C. O. Tichenor
Gardiner Lathrop

PART ONE

While the congressional party, with Major Gee, were making an inspection tour on the government snag boat, from St. Joseph to St. Louis, in May, the writer was reminded of a trip made down the river during the exciting days of the Civil War, and believing a few reminiscent articles would be of interest, we herein give an account of a trip made on one of the best known steamboats of that day.

Martial Law in Kansas City

As an introductory, a few thoughts showing the conditions existing in and around Kansas City will be necessary to explain the exodus which crowded the boats of that time to their full capacity. The historical fact well known as General Order No. 11, issued by Maj. Gen. Thomas Ewing, then stationed at Kansas City, in which every family with Southern sympathy was driven from home, in the counties of Jackson, Bates, Cass and part of Vernon, with a few exceptions in and around Kansas City and Independence, had been in force from August 25, 1863. The causes leading up to this, spoken of by some as the most outrageous and infamous order issued by any general during the Civil War, had been of long duration, known as the border warfare, raging for years between the fighting factions in Kansas and Missouri, known as Jayhawkers and Redlegs of Kansas, and the Southern sympathizers in Missouri who were driven to the brush and known as Bushwhackers.

The frequent encounters of these two opposing factions, threatening the peaceful attitude of the law-abiding citizens for years, engendering a revengeful spirit, culminating in the climax when a stern-wheel steamer, the Sam Gaty, was captured at Sibley Landing by Captain Gregg and his party of some eight to ten men. This boat was loaded with government stores, and while it was not burned, the government cargo was taken and appropriated. This called for extreme retaliatory measures to be taken by the federal authorities at Kansas City, and about this time Miss Jennie Anderson, a sister of Bill Anderson, and three other girls who espoused the cause of the Confederates, were taken prisoners and placed in an old condemned building on Grand avenue near 3rd street, which collapsed, killing all four of them, and Bill Anderson, one of Quantrell's band, swearing vengeance, led them to Lawrence, pillaged and burned the town, when it is said that more than one hundred were killed and the financial loss estimated to be near $2,000,000.

Order No. 11 Bingham's Painting

The federal troops of Kansas City were never able to find and capture these men, and these extreme atrocities led to the issue of Order No. 11, made so familiar to the people of Kansas City and Missouri by the graphic scenes depicted in the great painting executed by Maj. George C. Bingham of Kansas City. It has been said that the wife of Major Bingham was the first white child born in Kansas City, and with many other pioneers her grave is found in Union cemetery, for which many of our public spirited citizens are now fighting to retain intact as a part of our public park system, that the memory of our founders may not be lost.

That order was issued August 25, 1863. and that painting portrays the scenes and desolation as its stern decrees were enforced depopulating more than three counties in fifteen days. These people, the very best of Western Missouri, had to make their way as best they could, some into nearby counties, some going as fas as St. Louis and Illinois. In the surrounding counties they were known as refugees from Jackson County.

Steamboats Crowded With Jackson County Refugees

The steamboats then navigating the whirling eddying waters of the "Big Muddy" were crowded to their overflowing capacity from that time on to the close of navigation both in the fall of 1863 and 1864, and our old-timers can recall the names of many of these famous boats, such as the Polar Star, the War Eagle, the William J. Lewis, the Sioux City, and many others we could name. It is of a trip the writer made on the Sioux City in November, 1864, of which we write. She was a large side-wheeler running as far up as St. Joseph, under the command of Capt. Charles Baker, with Capt. Frank Woolfork at the wheel, two of that type of fine men who made steamboating a pleasure before and during the days of the Civil War and up to the 'seventies, when the railroads made such inroads upon their carrying trade that they soon went into decline. The old pilots of those days knew of the ever changing currents of the river, and kept well posted as to the bars and snags, which ever opposed any napping on the job, for each bend changed the course of the boat from first one side near the bank to the opposite side, so that many extra miles were traveled between St. Joseph and St. Louis, perhaps six hundred miles in low water. This was long before any appropriations by congress were thought of, or hoped for, and the pork barrel had not been emptied, as in the present day, without any appreciable difference in the flow of the mighty river. While it was not written up in those days as Mark Twain wrote of his pilot days in "Life on the Mississippi," yet the pleasure of a ride upon its unruffled swirling waters was none the less a delight, and was not taken without a desire to repeat the invigorating trip, with its quiet gliding down near the banks, with the mighty forests which lined it at that time, overspreading trees casting shadows on its warm decks in warm weather, making every day an enjoyable one as the passengers sat upon the cabin deck, touched by the gentle breeze as the fast receding trees and objects on the bank seem swiftly gliding by.

Mark Twain On a Pilot's Knowledge

We advise the reader to take down Mark Twain's "Life on the Mississippi," and when enraptured with his description of the ever changing channels in the Father of Waters, simply enlarge that about seven-fold to give a correct idea of the idiocyncracies confronting a pilot on a Missouri River steamboat. Mark Twain's summing up of what a pilot should know, is thus given:

"Two things seemed pretty apparent to me, one was that in order to be a pilot a man had to learn more than any one man ought to be allowed to know; and the other was, he must learn it all over again in a different way every twenty-four hours."

A Tribute to the Missouri

Remember, the Missouri River from its head waters in the Rocky Mountains to the sea is the longest river in the world, near 4,300 miles, thus exceeding the Amazon more than 600 miles. Having made several trips on the Missouri in its prime steamboating days, we gladly espouse its cause and welcome the day when it stands as the great equalizer of proper and just freight rates, from the great central agricultural belt to the markets of the world, and the farmers and producers will receive their equitable share of the products of the richest valley of the world.

One trip in 1868 was made on the Post Boy, which at the time carried the Horns, a very fine pair of antlers, golden mounted, a trophy won on a previous trip from St. Louis to Kansas City and return, in about five and one-half days.

Greatest Steamboat Race on Record

In 1853 the steamer Polar Star made a remarkable run from St. Louis to St. Joseph in 64 hours, and in July, 1856, the James H. Lucas made this run in 60 hours and 57 minutes. There are no recorded historical steamboat races on the Missouri in its prime days, but the races on the Mississippi from New Orleans to St. Louis have been of world-wide interest. The greatest race on record was between the R. E. Lee and the Natchez, on June 30, 1870, the Lee winning the race in three days 18 hour and 14 minutes, the Natchez arriving six hours and 36 minutes later. The ordinary run of boats made a speed of ten miles an hour, but some have made as high as fifteen miles, while some were very slow.

Uncertainty of River Transportation

For instance, the writer once came to Kansas City to engage shipment of hogs to St. Louis. He passed the boat at Missouri City, on the early morning Wabash train, and had to lay over until next morning. While this boat was slow, it was probably a champion racer by the side of the John J. Roe, as described by Mark Twain as being "so slow when she finally sunk in Madrid Bend, it was five years before the owner heard of it." It was customary for boats to tie up during the night, in traveling up stream, except during the spring rise, or in high water season, for the reason that the snags, on account of the heavy butts of the trees lodging on the sand bars, becoming imbedded, the jagged pointed limbs swinging around and pointing down stream, would tear a hole in the hull, upstream, but downstream, the boat would glide over or slide off these snags.

A Small Boy's Impression of Steamboats

The writer was but a boy, and his father in St. Louis had seen Capt. Charles Baker and made reserv ations for his family to be brought from Hill's Landing in Carroll County to St. Louis, in November, 1864, and we arrived at the landing ready for our departure on Thursday. The delay in the boat's arrival gave an opportunity to watch the swirling waters undermining the alluvial soil of that fine bottom land, as it caved in in great chunks daily, which has continued from that day to this. Carrollton at that time was reckoned some eleven to twelve miles from the landing, but at this time only about seven miles from the river, thus showing you where we boarded the Sioux City in 1864 has long since been transferred to the south side of the river. Many thousands of acres of the finest bottom lands in Missouri are thus destroyed annually for lack of proper check to keep the channel from carrying down everything before it wnen striking in full force in these great bends.

Starting of the Sioux City

The Sious City was probably having her troubles before reaching Kansas City, for she was

more than four days behind her schedule in reaching our landing. From Kansas City, where she received a good passenger list with much household goods stacked all around the cabin deck, she begins her voyage, probably making her first stop at Liberty Landing, where part of the belated hemp crop of the year before was added to her tonnage; then touching at Missouri City, Napoleon, and wood landings at various points, she is soon tied up at Lexington, then as now considered as one of the most cultured towns on the Missouri River, and in earlier days, possibly the third largest town in the state, antedating Independence and Kansas City many years. Even at this time Lexington was talking of bridging the Missouri River, and continued talking bridge until now its hopes are soon to be realized in the bridge now building. In those days the Missouri River held complete control of this part of the state as the Hannibal & St. Joe railroad was the only line then reaching its western border at St. Joseph.

Dancing on the Old Steamboats

At Lexington, Dover and the small villages then lining the river, it was customary for the gay young parties to board these river steamers, and after the evening meal had been served, the tables and chairs removed from the cabin salon, the music maker, at that time an old time fiddler, soon tuned the strings and rosined the bow, and the mating of gallant young men and your grandmothers, in the styles of that day, with graceful courtesies, saluted their partners and were ready to execute the most intricate evolutions in the Virginia reel or other square dances of that period. When you look upon the styles of your grandmothers, in comparison with your own untrammeled skirts, you no doubt wonder that it was possible to "trip the light fantastic toe" in that day, with the grace required of an accomplished dancer of today; but your fathers. and grandfathers have often looked upon them with pride as they not only executed the many whirls with the spreading hoopskirts, some eighteen to twenty feet in circumference, but in addition have seen them carry from three to five feet in trains to the gowns of that day.

While the cry was "On with the dance," the boat was fast approaching one of the most historic of old river towns, and it took but a few minutes for these revelers to persuade the captain, if the night was moonless, to tie up at Waverly, Mo., and the gay young people were in for an all-night dance.

Old Waverly Home of Gen. Joe. O. Shelby

Waverly in the early day before the Civil War, was the home of Gen. Joe O. Shelby, and his home, on the bank of the river, just under the bluffs, with its dormer windows, and the old rope walk or factory, were for many years pointed to with pride by the dwellers in the grand old Missouri River town. We had hoped to give you pictures of this town in its prime before the Civil War, but the present day kodaks were not functioning. Most of the fine people who inhabited this part of the state and made Waverly a great shipping town in the hemp-raising days before the war, have scattered. General Shelby and many others in Waverly were hemp raisers before the war, and being of that extreme hospitality characteristic of the early settlers, those who had come to Missouri from South of Mason and Dixon's line, principally from Virginia and Kentucky, these people joining in with the others up the river, kept that boat in a quiver until the approaching dawn compelled the captain to call a halt, and the young people took the stage back to their homes or caught the next boat going up the river. Waverly is now a railroad town, but the great bridge across the river, now under construction, puts it on the map as one of the strategical points on one of Missouri's new lateral highways linking the north side of the state to the south side, thus welding it in a commercial way as well as politically, as it has never been hitherto.

There were many men who lived in old Waverly whose recollections of the old war days, and those who followed Joe Shelby through the war and down into old Mexico, of whom that brilliant writer, John N. Edwards, made heroes, in his book, "Shelby and His Men," and there are today ladies living in Kansas City who have vivid recollections of a certain episode transacted near Waverly in which two of General Shelby's captains took the leading part.

General Shelby and Duel at Waverly

The valor and chivalry of the old South still trickled in their veins, and Captain R. C., for some trivial misunderstanding in which two of Waverly's most charming daughters were involved, felt that his pride required him to issue a challenge to Captain W. R., one of his very best friends, and General Shelby was made referee in the affair, choosing the weapons to be used, and it was due to his quick and ready forethought that their question of Southern honor was forgotten and they lived in old Waverly for many years as good friends. Captain C. and Captain R. were both welcome in one of the noted homes of old Waverly, where the hospitality of those days was at its height, and the attraction of several beautiful daughters was sufficient to keep not only the young men of their home town careful and watchful of their behavior when these ladies appeared on the streets, but also attentions were paid them by distinguished visitors from other parts of the state. We have not at this date the full particulars leading up to this challenge, but recently met the two ladies who were deeply interested in the outcome. The challenge was issued by Captain C. and was instantly accepted by Captain R., and General Shelby selected those faithful old weapons that had been the most reliable to him in many a skirmish, the old double barreled shotgun which he loaded with buckshot, and upon arriving at the scene on the field of honor near the old covered bridge, at sunrise on the appointed day, after the goodbys had all been said, General Shelby called both the participants before him, presenting to each of them a gun, but cautioning them as to how he had loaded them, believing that if the triggers were ever touched each of them would be sent to the Happy Hunting Grounds, without any question, and speaking the right word in due season, they shook hands, and we are certain the young ladies were agreeably surprised at the outcome. Both married in after years, but neither to the young ladies for whom they were willing to pour out their life blood.

Leaving the Sioux City tied up at old Waverly for the night, we will be ready to board her the next morning to continue the memorable voyage.

PART TWO

Goodbye to Old Waverly

At break of day the bell is sounded, the roustabouts, negro deckhands, heave-ho with the great stage plank, the hauser is pulled in, Captain Woolford is at the wheel and the Soux City slides out from her moorings, turning to the right her bow is soon pointing southeast as she enters midstream feeling for the current which bears to the east about two miles, to the Cottonwood landing, where she ties up to supply herself with the needed cordwood piled up for the river trade, and while the mate hustles his routabouts up and down the stage plank in a run, we will give a few episodes which are vivid in our memory, before bidding a final goodbye to old Waverly.

A Ducking in Icy Waters

The father of the writer, a dry goods merchant in Carrollton before the war, having some business in Waverly in the early spring of 1860, rode from Carrollton on horseback in the early morning, and, leaving his horse on the north side of the river, crossed the river on the ice, being cautious he provided himself with a long cotton wood pole. Having completed his business, a bright sunshiny day had effected the ice until some of it was moving, when he came down to cross in the late afternoon, and when possibly two-thirds of the way across, down he went into the icy waters, but with the assistance of the pole he managed to gain the solid ice again, and with frozen clothes made his way on his fleet horse to his home. The motion of his swift horse soon covered the 11 miles and he was soon made comfortable and a very thankful man.

Steamer Sioux City on Her Last Voyage, 1864

Gen. Albert Sydney Johnston

About the same year, 1859 or '60 one of the mountain boats sank just above Waverly, upon which Col. Albert Sydney Johnston, then returning from Fort Benton, was a passenger, and escaping to land spent a night in Waverly. Gen. Albert Sydney Johnston was afterward one of the most distinguished Confederate generals during the Civil War. He took the next boat to St. Louis.

On Sunday morning, while playing on the bank of the river about two miles up the stream around the bend in the willows, we saw issuing from the steam whistle that column of white steam so visible from an approaching boat, and thinking it the long delayed Sioux City, we soon had our mother and the family assembled on the bank, but it was but a few minutes until we discovered our mistake, as the trim little stern wheel steamer, the Hattie May, hove in sight, and tying up a few minutes, was soon gracefully gliding downstream, now drawing not more than three feet of water had no difficulty in navigating over its bars. Tuesday morning, after the mate had supplied himself with several cords of that light cottonwood wood, the boat was again on its way and soon we heard the well-known whistle as it gave us the landing warning, a long and two shorts, a long and two shorts, as she rounded the bend swinging into midstream, past old Buffington landing, upon whose banks the historic old mansion stood, and nearby the homes of Capt. Charles Baker and Captain Woolfork, and also Capt William Baker, for these steamboat men had settled in what was known as the sugar tree bottom of Carroll County. As the boat approached, swinging down stream, the famous name "Sioux City" was plainly visible in large black letters, over the covering of the great side wheels, and soon she turned headed up stream and gracefully gliding into her berth at the landing, when two of the most agile roustabouts scrambled up the twenty-foot embankment with the great three-inch hauser, to make her fast to the bank, and at the same time some dozen or more were pushing the heavy stage plank to the top of that bank, and in a few moments we were on board and assigned to our comfortable staterooms, which had been reserved for us on a very crowded boat. A few hogsheads of tobacco were soon rolled down the gangplank, the bell sounded, and we were off for St. Louis.

Off For St. Louis

Seeking the middle of the channel, as the Big Muddy, then very low, continued its tortuous course, now through some long chute near its bank where the great overhanging trees often swept near the deck, then again threading our way between the sandbars and the ever invisible snags, sometimes scraping the bottom. The Hattie May made progress where our boat, loaded down to about four feet, had great difficulty in getting over many of the bars. The frequent taps of the bell calling for the heaving of the lead, became very familiar to us, as well as the well-known calls described in Mark Twain's book, as the lead heaver called out "f-i-v-e feet" and up to nine and a half feet, then "m-a-r-k t-w-a-i-n" and "q-u-a-r-t-e-r-l-e-s-s t-w-a-i-n," when the bell was again tapped and the lead was called in. Often in touching the ledge of a bar he would sing out "t-h-r-e-e feet." The lead heavers were quite expert in their calling and were specialists in their line. We were told of an amusing incident on one of the boats, in which the regular lead heaver was unavoidably off duty and a new man, of German proclivities, was ordered by the mate to handle the line. Not knowing any of the terms, he cast out the line and sang out "much vater," and shortly afterwards, approaching shallow water, he called out"not much vater," and in a moment more the boat was on the bar, and when reprimanded by the mate he replied, "I told you there was not much vater."

Our first landing was Malta Bend, Saline County, where we took on a heavy shipment of wheat from the celebrated Pteetsaw plains in Saline County, near Malta Bend, noted before the war as rich hemp lands, which in its prime days has probably produced its fifty bushels of wheat to the acre, and in the Indian days, grass—the wild prairie grass, grew as high as a man's head on horseback. These lands are said to be equal to the best in all Missouri, and hence unsurpassed in the United States.

Miami a Noted River Town in Early Day

Our next stop on the Saline County side, is Miami, a town situated high and dry, on a bluff, settled in a very early day by the Indians, the Miamis, from which it takes its name. Like Waverly, Miami was one of the oldest and most prosperous of the river towns, being settled back in the thirties, was in its prime three times as large as Marshall, the county seat. The settlement was first called Greenville, but changed to Miami in 1837. Saline County, whose boundary both on the north and east is the Missouri River for many miles, the largest county in the state, has been known for many years as the great corn country, and this county produces more corn than all the New England states. Miami in early steamboat days was the home of many steamboat captains, among them Captain Burress and other noted river men. Captain Burress died last spring at Carrollton, Mo., somewhere near the ninety mark.

The Beginning of Prof. John T. Buchanan's Brilliant Career

It was in this town of Miami where our distinguished townsman and educator, Prof. John T. Buchanan, began his brilliant career, as principal of the school, afterwards for ten years principal of Central High School of Kansas City, and from there to New York, where the Board of Education erected a school after his own design and placed under his leadership—the Boys' High School—until his death, some fifteen years ago.

The next town we touch on the north side of the river is DeWitt, Carroll County, quite a stopping point in steamboating days, and a transfer point in after years when the Wabash road put in a boat, the A. C. Bird, to transfer grain from the rich lands of Saline to the transfer point at DeWitt. DeWitt figured conspicuously in a very early day as the central point of Mormon history in Carroll County, and in passing this quaint old Missouri River town a few notes may be interestng.

The Mormon War in Missouri

The Mormons first settled in Independence in 1832, but in 1834 were driven out of Jackson County, and settled in Clay, Caldwell, Carroll and other counties, making headquarters at Far West in Caldwell County. In 1838 two of their leaders settled in DeWitt, which at that time they considered a good shipping point for Far West and about the head of navigation on the Missouri River. Colonel Hinkle and his associates soon became obnoxious to the people and they ordered him to move on, but he threatened extermination to any who molested his Mormon recruits, who, gathering there from all directions, camped in great numbers. Gentiles, as those oposed to them were called, one hundred and fifty strong, also camped near DeWitt—a slight conflict occurred September 21, 1838, but nothing serious resulted from this first encounter. The Mormons evacuated and fled to some log houses, while the Gentiles awaited reinforcements, who came from Ray, Saline and other counties, five hundred in number. Congreve Jackson was chosen Brigadier-General and Ebenezer Price, Colonel, and after a few days drilling were prepared for an attack, but before commencing action, Judge James Earickson and William F. Dunnica from Howard County adjusted the trouble without bloodshed, and the next morning the Mormons began their move to Far West. General Doniphan was sent to the seat of war by the governor, and General John B. Clark was placed in command of the militia or state troops. Although the Mormons had left Carroll County, they continued to augment their forces on their march to Far West, and were one thousand strong when a slight skirmish occurred at Crooked River in Ray County, in which one Mormon was killed, but later the principal ight was at Haughman's Mill, where eighteen Mormons were killed and the balance captured, some of them being killed after surrendering. General Doniphan finally captured Far West, and this closed the Mormon war in Missouri.

Another Famous River Town—Brunswick

The Sioux City, after picking up all the tobacco, hemp and wheat at these points, is again on her way, passing the mouth of the Grand River, the principal tributary of the

Missouri River, from the north side, we round to a mile below, tying up on the east side at one of the most flourishing towns of that day, Brunswick, in Chariton County, the leading shipping point and tobacco center of the whole state, at that time, now upon the Wabash railroad and not considered on the maps as a river town, the Missouri River having cut across into Saline County, leaving Brunswick many miles from the main river, the Grand River now flowing by it in the old Missouri River bed.

Some Episodes in Our Boyhood Days

We well remember two episodes in our early life in which Brunswick was closely related, one occurring a few months before this trip on the boat, and the other a few years later, in 1867. In the fall of 1864 there was opened at Glasgow, Mo., two schools, the best in this part of the state at that time, the one for young men, known as Davies Academy, the other for young ladies, known as Strother's Academy for Young Ladies. My father sent an older brother and a cousin to the Davies Academy, and an older sister to the Strother school, and one of our near neighbors sent a son also to Davies. After Price's raid through Missouri that fall, the mother of this neighbor's son and my aunt, becoming anxious about the boys and not hearing from them, concluded to visit Glasgow and find out, and hitching the old family buggy horse, old Charley, to the surrey, and taking the writer along, we were off on the trip. The first day out of Carrollton, as we were nearing Brunswick, in passing through that great hickory timber along the Grand River bottom, we met Bill Anderson and his full company of the most noted Guerilla band in Missouri, and driving to one side of the road to watch

them pass, an officer rode out of the ranks, and informing us who they were, pointed out Bill Anderson riding about the middle of the company. We were soon ferried across the river in the old-time flatboat, pulled by ropes, and in Brunswick for the night.

The other episode in which Brunswick will always be remembered was in the fall of 1867, when a younger sister and the writer made the trip by stage from Carrollton to old Allen, now Moberly, Mo., on the old North Missouri Railroad. We started from Brunswick in the stage hack early in the morning, but before 10 a. m. one of Missouri's phenomenal rains began, the poor beasts gave out and we were left upon the prairie in that storm while the driver rode one of the horses into old Fort Henry, near Salisbury, and got a fresh horse to bring us into Huntsville, where we transferred to one of those old-time Rockaway stages, such as you may see in a Wild West show at present, and soon were in old Allen, but it being Saturday evening, we learned there would be no passenger train that night, as passenger trains did not run on Sundays in those days. An accommodation freight with a passenger coach soon appeared, and we made St. Charles by breakfast the next morning. In these early days Brunswick was the principal shipping point for the tobacco raised in Chariton County, and the great tobacco factories were very conspicuous. We are now in the very heart of Missouri, and Chariton County holds the record of being the largest tobacco raising county in the state, and the highest record ever made by any county, and therefore in the world, was given it when in one of its best years, it produced thirteen million pounds. The Sioux City was here loaded down with many hogsheads of tobacco, and in the passenger list a woman and five children with Scarlet fever were taken aboard. Resuming our journey on the Sioux City, not a pleasant one to contemplate, in a crowded boat and five children with Scarlet fever, which quickly spread to other children on the boat, and the writer now has a sister who lost her hearing from that experience.

Old Keytesville Landing Only a Memory

Our next landing was Keytesville Landing, at that time one of the distinguished landings along the river, for the Missouri River here touches the two banner counties of the state, Saline and Chariton, old Keytesville, the county seat of Chariton, being some eight to ten miles from the river in the very heart of this old county wherein much of the early history of Missouri was made. From Keytesville Landing back to the bluffs and stretching along the river some twenty miles or more is what is known as Bowling Green Prairie, known all over the state for the richness of its soil, where in ante bellam days the country was settled by descendants of the Virginias and Kentucky from which are numbered one of the most distinguished of all Missouri's great men, that most noted western Confederate general, Sterling Price, whose home was but a few miles from this landing, surrounded by a farm of some eight hundred acres, in his day one of the great hemp raisers of the state; but when his country called he was among the first to respond, and possibly won the last battle of the Mexican War in 1848. This old county and its adjoining county, Howard, are so full of thrilling incidents in the early days of our state's history that we have determined to leave the memorial erected by our state to its distinguished men, for a future article, for when a writer steps upon the sidewalks of these old Missouri towns and meets the descendants of those who paved the way, who cut their way into its great forests, he must have space to do it justice. Our notebook is full, but we haul in the gang plank, and again our boat is headed down stream; but we will note in leaving the old landing, that its historic old warehouse, stores and great trees, have long since been swept away, as well as thousands of acres of this alluvial soil, now covered by willows on the Saline County side. From Brunswick to Glasgow, a distance of some forty miles, Chariton has lost probably four thousand acres of her best lands, added to Saline County, and possibly gained an equal quantity at other points. Chariton County is well watered by the east and west forks of the Chariton River. As we near Glasgow, we pass the mouth of these two rivers, which, when discovered by Lewis and Clark in 1804, emptied into the Missouri as two distinct streams, but now as one. The boat has passed old Cambridge, on the Saline County side, and as the evening sun sets in the west, her whistle blows the long, two shorts, and the long and two shorts, and we tie up at old Glasgow for the night.

PART THREE

When the boat tied up at Glasgow in November, 1864, we were at a landing known from the earliest boating days of the Missouri River as a good landing for steamboats, and now since more than a hundred years has passed Glasgow is still a good boat landing, so far as the wharf is concerned, as no part of its wharf has been washed away, nor has the river ever attempted to wander off in some new channel. Perhaps this is why the Chicago & Alton railroad and the state highway men have chosen it as a site for the two bridges, one dedicated on June 6, 1879, the other road bridge now under construction.

Battle of Glasgow

As the boat tied up at this historic old town, it was not long before Professor Strother escorted several of the young ladies of his academy down to see it, and among the number was an older sister of the writer. As Price had made his second raid through Missouri only one month before, and the battle of Glasgow was still fresh in the memory of the young ladies from Strother's Academy, we will sketch briefly the bit of history next to the capture of Lexington on his first raid, the most important engagement of the advance in Missouri, in that year. We give the date of that battle as October 15, 1864, but find in the different accounts now on record that these dates are given October 10th, 15th and 20th. General Joe O. Shelby ordered Captain Collins' battery to open fire on the town from the Saline County side of the river, just at sunrise, and Shelby, taking Strother's Academy for the City Hall, in which he supposed the garrison was located, had directed Collins to fire at the Academy, and the first or second shot knocked off the northwest corner of the building. Professor Strother aroused his students, (and those girls probably dressed a little quicker than they had ever done before or have done since) and marshalling them in front of the building showed General Shelby his mistake, and he changed the direction of the fire to the City Hall. Major Pratt's battery approached on the Howard County side, captured a boat, the Western Wind, and then the City Hall. Col. Chester Harding was in intrenchments on top of the hill, south of Strother's Academy, with the 43rd regiment Missouri Volunteers, Gen. John B. Clark, jr., approached the town from the Fayette road, over which Professor Strother marched his pupils to safety beyond the town. The City Hall and about a dozen houses were burned, and the garrison surrendered about 1 p. m. There were some fifty to sixty men killed and wounded on each side. The writer passed through the temporary hospital, one of the churches, the only one he ever visited during the war. The City Hall was still smoking when we arrived, searching for the boys at Davies' Academy, but both had gone with Price.

Outfitting Point For 49er's

Glasgow was one of the oldest shipping points on the river, at one time, like all of these old Missouri River points during the excitement of the forty-niners, being an outfitting point for the "Santa Fe Trail," but there were other historic towns on the Chariton River, just above Glasgow, that antedated the founding of Glasgow by some twenty years. Lewis and Clark's great exploring expedition which went up the river in 1804, speaks of passing Deer Creek, and five miles further the two rivers called by French "Charitons," a corruption of Thieraton, first thirty yards wide, and second seventy-five yards wide. French and Indians had settled at old Chariton prior to 1804, but the first white man of whom there is any record before 1812 was George Jackson, in the southern part of the county near the Missouri River.

Old Chariton Laid Out in 1817

Old Chariton townsite was laid out in 1817. James Earickson settled in the Missouri River bottom west of Old Chariton, and was afterwards a state senator and treasurer of the state. Many good men soon settled in these bottoms, as far north as Bowling Green Prairie. Indians were here prior to all settlement by the whites. The first steam mill was erected in 1817. Old Chariton was the most thriving point beyond St. Charles, in the state, and as it became the fartherest point of navigation at that time, all outfitters started from here to the great Northwest. Excitement became high—something like the oil towns of today, until many expected it to rival St. Louis—in fact, lots in St. Louis were exchanged for lots in Old Chariton.

Once a Rival of St. Louis

William Cabean, a shoemaker in St. Louis, sold his property there, about a block from the courthouse, for $3,000 and invested in Old Chariton. The St. Louis property is now beyond the $100,000 mark, while Old Chariton is but a dream of the past. The Chariton River was once considered navigable for steamboats. Old Chariton reached its highest mark about 1826—about two thousand inhabitants. The floods and consequent sickness drove them to higher ground, and old Monticella was founded in 1831, and finally Glasgow in 1837, named for James Glasgow, one of the early settlers.

First Mail Contract in Western Missouri

The first mail contract let by the government, west of old Chariton, was in 1833. James Wilson was the first mail contractor, and his son carried the mail on horseback from old Chariton to Liberty, Mo. Charles Mann was second, then John M. Davis succeeded him October 9, 1833, who was then a boy of fifteen years. The route was 130 miles, starting from old Chariton to Keytesville on Wednesday, and then to Grand River, where he spent the night, and the next morning to Carey's postoffice in Carroll County, then to Henry Bresden's ferry on Crooked River, where he spent the second night, and breakfasted at Richmond, Mo., reaching Liberty that evening, returning to old Chariton, arriving in the afternoon of Monday. His pay was $9.00 per month, and his board and expenses, and he provided his own pony to ride. The westbound mail was about three pecks, but the eastbound could be put in a stovepipe hat. Davis was at that time a physically weak boy, but afterwards became a judge, and resided for many years at Brunswick. It is said there was only one German in Chariton County from 1820 to 1835.

Methodist Preacher Founder of Keytesville and Brunswick

Keytesville, the county seat of this county, was laid out in 1832 by James Keytes, a Methodist preacher, who afterwards in 1836 laid out Brunswick and was its first postmaster, and continued in the office until his death in 1844.

While we have given some historical facts of the early settlements of Chariton County, Glasgow is in the very corner of Howard County, and the history of the two is so interwoven it is difficult to separate them in their early history, and before we resume our journey we will give the record of the port of Glasgow in 1849 and 1850, showing very conclusively just what was being done in the prime shipping days, by water.

In 1849, there was shipped from the port of Glasgow, by boat, tobacco 5230 hogsheads; hemp 3577 bales, bacon 118 casks, coils of baled rope 1250, lard 259 barrels, lard kegs 320, green apples 4471 barrels, dried apples 4089 bushels, wheat 21,670 bushels, dry hides 935, pork 450 barrels.

Port of Glasgow, 1850

Arrived from St. Louis

Sacramento	April 19
St. Paul	April 19
Lightfoot	April 21
Monroe	April 21
L. L. McLean	April 21
General Lane	April 22
Minnesota	April 22
El Paso	April 22
Pocahontas	April 23
Tuscumbia	April 25

Departed for St. Louis

Mary Blane,	April 18
Haydee	April 20
Jas. Mullinger	April 20
Hungarian	April 20
St. Ance	April 21
Princeton	April 21
Alton	April 22
Cambria	April 22
Robert Campbell	April 22
General Lane	April 23
Ne Plus Ultra	April 23

Population of Glasgow in 1852, 1,000.

Dedication of First Steel Bridge in World

Glasgow continued to be a river town until the building of the first great steel bridge in the world, across the Missouri River, by the Chicago & Alton Railroad, which was dedicated June 6, 1879. The writer of this was present on that occasion, when the first great all-steel bridge in the world was tested by passing over it eleven heavy locomotives, and the test was satisfactory to the engineers. Dr. Pope Yeaman, one of the celebrated orators of the day, delivered the address to seven thousand people who had been brought in from all along the line by the C. &. A. and every known vehicle from the adjoining counties, possibly the record picnic of this Missouri River town. In his address, he related an incident given him by General Smith, the engineer who built the bridge, in which General Smith, while in London some time before, met the celebrated English engineer, Mr. Barlow, saying to him, "Look out, Mr. Barlow, or we will build a great steel bridge in America before you will in Europe."

A few days ago, Dr. Yeaman said, General Smith received a letter from Mr. Barlow asking how the proposed bridge was progressing. The general, with great satisfaction, replied, "Trains are crossing it." The Morrison observatory is located at Glasgow.

PART FOUR

Our stay at Glasgow was prolonged on account of the very low stage of water, the captain preferring to take more time, as it was his last trip for the season, than risk his boat and cargo at night.

Anti-Prohibition Days—"The Thrill of the Trip"

We were now well loaded as to tonnage, and also a very unwelcome passenger came aboard here, for he, too, was well loaded. This man was a soldier from a company of Missouri volunteers known as Merrill's Horse, and as prohibition was not in force, he gave us the greatest thrill of the entire trip, for unloosing from Glasgow at an early hour, we were soon approaching one of the most noted historic towns along the Missouri River, old Arrow Rock, but just before reaching it and while gliding smoothly down near the Saline County bank, this soldier, just sufficiently under the influence of liquor to stand upon his feet, yet knowing enough to make him feel brave, seeing several men on a high bluff from a quarter to a half mile away, he stepped over the guard rail on the cabin deck near the front end of the boat and emptied his revolver at them, and they, quickly accepting the challenge, returned the fire with their rifles. The bells were rung for full steam ahead, and Captain Woolfork lost no time at the wheel in changing that boat for the distant shore, while Captain Baker and his efficient clerk ordered the women and children to drop on the floor inside the cabin, and crouch down behind the walls of the staterooms. The minnie balls flew thick for a few moments, some striking the pilot house and smokestacks, but fortunately no one was hurt, but a week or more after, we read of the death of that member of Merrill's Horse, near Arrow Rock, where he left the boat. The captain touched at Arrow Rock for only a few minutes, but about three miles below, the boat landed on a bar, and as we remained there three days and nights, we will let the Sioux City rest while we relate something of this historic old river town.

Old Arrow Rock

It has been written up time and again, and just recently at the celebration, where the state bought the old mansion or inn, in which many

of the noted men of the steamboating days were wont to gather. We have all heard of the stories of Claiborne F. Jackson, and the three wives who came from one family in Arrow Rock, and the story of Dr. John G. Sappington and the foundation of his fortune, made from peddling his pills, in the early days, when it is said he has ridden on horseback from Kalamazoo, Mich. to New Orleans, so that he did his own advertising of Sappington's pills. Dr. Sappington provided in his will a fund for indigent boys, known as the Sappington School Fund. The origin of the name Arrow Rock has been given several interpretations, but possibly the best explanation is the one associating the large rock on the river bank as a mark for the practice of the Indians, in the earliest days, shooting their arrows from the Howard County side of the river, at this rock.

Some Early Howard County History

While the boat still stands on the bar below Arrow Rock on the Saline County side, it gives us time to reminisce on the Howard County side, where we find the historic town of Old Franklin, possibly the very earliest settlement of any importance outside of St. Louis and St. Charles, in the state. In 1807, Nathan and Daniel Boone, sons of Daniel Boone, the trapper, went from St. Louis to Boone's Lick in Howard County, and with three others, commenced the manufacture of salt, forming the first white settlement beyond St. Charles, and in 1810 a colony of 150 Kentuckians came to Howard County, settling at Old Franklin, opposite Arrow Rock. The first steamboat on the upper Mississippi landed at St. Louis in the year of 1817, which was the General Pike, and the first steamboat up the Missouri River as high as Old Franklin was the Independence, in May, 1819. Lewis and Clark's exploring expedition had passed up the river in 1804. Howard County was organized from St. Louis and St. Charles counties in 1815, and its boundary was thus described: "All territory north of the Osage River and the dividing line or ridge between the Mississippi and Missouri rivers."

The first grist mill was set up at Old Franklin in 1819, cost $50.00, capacity 50 bushels per day. People came from near and far, camping out awaiting their turn at the grist. A traveler at this time saw 75 deer in a distance of six miles. There were many bee trees, and honey commanded a good price. It is supposed Daniel Boone camped at old Boone's Lick in some of his early hunting expeditions about 1800. The salt manufactured by his sons in 1807 brought about 50 cents per bushel. Col. Benjamin Cooper came to Boone's Lick in 1808, but the government, recognizing his encroachment upon Indian lands, had him withdraw to a point below the Gasconade, and he settled Loutre Island, but returned to Boone's Lick in 1810 with some twenty-five families, when they erected cabins and put in crops.

Ira P. Nash obtained a Spanish grant, which settlers held and Indians acknowledged, in 1800. These people manufactured their powder and salt and were plentifully supplied with wild game meat. They supplied themselves with a fabric made from wild nettles, as cotton goods. When the war of 1812 broke out, the Indians became hostile, incited by English sympathizers, so the people built Fort Cooper and Fort Hempstead in 1812, and all the settlers were kept in these forts until peace was declared. Judge Abiel Leonard taught school near Old Franklin, in 1819. He was afterwards Judge of the Supreme Court of Missouri. Robert Cooper and Elizabeth Carson married in the spring of 1810, the first marriage of record. She was a sister of Kit Carson, who was a son of Lindsay Carson, one of the early settlers. Kit Carson, the celebrated trapper and Indian scout, was born in Kentucky in 1809 and came to Old Franklin with his father in 1810. This family with others went into the third fort, Fort Kincaid, during the war. Kit left in 1826 for the Rocky Mountains, finally settling in Santa Fe, New Mexico, where he died in 1869. Three governors of Missouri came from Old Franklin, Lilburn W. Boggs, John Miller and Hamilton R. Gamble. It was the first outfitting point for the Santa Fe Trail. Claiborne F. Jackson, the war governor, was a clerk in a store there in 1826. Henry V. Bingham, the father of George C. Bingham, the noted artist, kept hotel in Old Franklin. The first three grist mills were built there, from 1820 to 1828. The Missouri Intelligencer, the first newspaper west of St. Louis, established in 1819. Travel to St. Louis was on horseback until 1820, when the first four-horse stage was run, and soon after steamboats with a $10.50 fare to St. Louis. Old Franklin was county seat until 1823, had from fifteen hundred to seventeen hundred population from 1823 to 1826, and was washed away by the river in 1828.

New Franklin was founded in 1828, two miles from the old town. In December, 1819, a letter from Old Franklin describing the first steamboat, says:

"In place of a bowsprit she carries a great serpent and as the steam escapes out of its mouth, it runs out a long tongue, to the horror of all Indians. They say, 'White man, bad man—keep greatspirit chained and build fire under it to make it run a boat.' "

Great Flood and Famous Duel

The great flood of 1844 lasted three weeks and water was six feet higher than ever known, and every building of these old towns was swept away. The falling stars of 1833 were observed by people in Howard County. Judge Abiel Leonard moved from Old Franklin in 1824, to Fayette, and in the same year fought a duel with Major Berry, an attorney of Old Franklin, on September 1, 1824, and as duels were prohibited in Missouri, this one was fought on Wolf's Island in the Mississippi River, thirty miles below Cairo. Berry was shot but not mortally wounded, but died a week after at New Madrid, of pneumonia from taking cold. Howard County has thirty-four miles of river front. Its population in 1860 was 15,946.

Some Early State History

In 1818 and 1819, the people made application for admittance to the Union, and the slave question began, known as the "Missouri question." The Missouri Compromise was introduced in congress by Henry Clay of Kentucky, by which Missouri was admitted into the Union, February 28, 1821, complete without any further debate. The legislature that met in St. Charles in June agreed to Clay's resolution, and August 10, 1821, President Monroe announced by proclamation the admission of Missouri.

Frozen in On Sand Bar

We have digressed somewhat while the old Sioux City is resting, three days and nights on the sand bar three miles below Arrow Rock. It was here we had the experience of seeing the great spar poles used to their fullest extent, and realized they were not strung up on the bows of boats for ornament, and noted the power of the capstans when the great cables were wrapped around them, and as they were drawn taut how the great boat was moved around until it stood broadside across the stream, and as the ice was running heavy, during the night it piled up against the boat, extending in a triangular shape for a hundred feet or more up the river, and packed tight, so that the stage planks were thrown out on it and plankways extended so the roustabouts could warm up as they attempted to cut us out. Their attempts were futile, and the Sioux City was thoroughly grounded and helpless, and might have been there yet if a rise of a foot in the river had not lifted us off the third night.

Boonville—Home of Steamboat Builders

Again we move on down stream, passing Boonville, where one of the first battles of the Civil War was fought, and where one of the great highway bridges of the state is now nearing completion. The M. K. & T. railway has a bridge at this point. As we pass along down the river, we are reminded that some of the noted boatmen of the Missouri River boating days, were brought up in the old Missouri River towns, includng Jefferson City, the Capital of our Commonwealth.

The Kate Kinney, the Joe Kinney, the Phil E. Chappell, and others bring to mind names that are ever linked with steamboating on the Missouri River.

Herman—The Impassable Bar

After this first rise in the river, the boat made fair progress for a day or two, until we reached the old Dutch town of Herman—when about three miles below, we found the river spread out, until there was no water that would carry us over, so we returned to Herman and unloaded the heavy freight, such as hogsheads of tobacco, and the next day again tried to cross the bar, without success, and we again returned to Herman, and unloaded practically all the freight on the boat, and again tried the bar, only to find the only hope the captain had of landing his over-crowded boat of passengers was impossible, so he again returned to Herman and tied the boat up awaiting a rise in the river, and sent the passengers on the Missouri Pacific railroad to St. Louis.

Crowded Train Without Water

If you ever rode on a train in those days, when a water boy passed water through in a tin pail, and heard the cries of those children with fevered brow, calling for water, as the scarlet fever patients suffered, you would realize how indellibly that trip on a Missouri River steamboat which failed to land us at our destination, has been stamped so vividly on the writer's mind. We arrived in St. Louis about midnight, and the Sioux City came down about two weeks later.

APPENDIX

List of steamboats on Missouri River from 1840 to 1852 furnished us by George P. Venable, Lexington, Mo.:

Arabia
Arotecis
Admiral
Anawan
Alton No. 2
Arkansas
Amavauth
Algoma
Amelia
Alice
Alton
Albatross
Anna
Andrew Jackson
Alex. Hamilton
Bowling Green
Balloon
Bertrand
Boonville
Boras No. 3
Bay State
Belle Creole
Bellow No. 2
Car of Commerce
Col. Wood
Columbiana
Chariot No. 2
Consignee
Cambria
Cora
Clipper
Cumberlain Valley
Courtland
Dianah
Dart
Duroc
Dakota
Daniel Boone
El Paso
Eagle
Emerald
Emma
Emory
Eliza Stewart
Frolic
F. X. Aubrey
Globe
General Brook
Henoplas
Huntsville
Henry Bay
Haydee
Highland Mary
Iatan
John Hancock
Jas. S. Lucas
Jas. S. McLean
John Nelson
John Golong
John Aull
Jewess
Julia
Jas. S. Hardin
Jas. Mulligan
Kate Kinney
Kit Carson
Kansas
*Mary Tompkins
Princeton
Pioneer
Plow Boy
†Polar Star
Rowena
Radner
Rhine
Robert Fulton
Robert Campbell
St. Cloud
St. Lawrence
St. Peters
Sawyer
Shawnee
Siam
Star of the West
St. Louis Oak
Silver
St. Joe
‡Saluda
Sacramento
San Francisco
St. Ange
St. Paul
St. Croix
Saranack
Sligo
Sioux City
Trenton
Trapper
Tippecanoe
**Twilight
Thames
Tobacco Plant
Tributary
Timonon
Tamerlain
Thos. Jefferson
Tuscumbia
Yellow Stone
Yucatan
Violet
Whirlwind
Wellington
Western Belle
Weston
War Eagle
Wyandotte
W. H. Russell
Warrior
Sky Lark
Thos. E. Tutt
Asa Wilgus
Swan City
A. B. Chambers
D. A. January
Minnehaha
Omaha
John D. Perry
Isabella

*Mr. Venable's family left Liberty and went to St. Louis on the Mary Thompkins in high water of 1844.

†Mr. Venable arrived at Lexington on the Polar Star, September 19, 1853. Still living at Lexington.

‡Saluda blew up at Lexington, April, 1852. One hundred lives lost. On May 17, 1849, twenty-three steamboats burned at St. Louis wharf.

**Twilight sunk at Lexington, Mo., 1852.

Steamer Alonzo Childs made the run from Kansas City to St. Louis between sunup and sundown during high water 1857 or 1858.

COLOPHON

This edition of C.P. Deatherage, *STEAMBOATING ON THE MISSOURI RIVER IN THE SIXTIES*, was printed in the workshop of Glen Adams located in Fairfield, Washington. Fairfield is on Highway 27 which runs between Opportunity and Tekoa. This edition of C.P. Deatherage was rephotographed from the previous 1971 printing. Additional typesetting for this edition was done by Teresa Ruggles. Dustin Newlun did the photography, film stripping, and design work. Printing was done by Trevor Del Medico using a Heidelberg, model KORS, printing press. Folding was done using a 26x40 Baum Dial-O-Matic folding machine. Paper binding was completed by Garry Adams using a Mark II Sulby binding machine. We had no special difficulty with the work. This is the 677th title to come from the Ye Galleon Press.